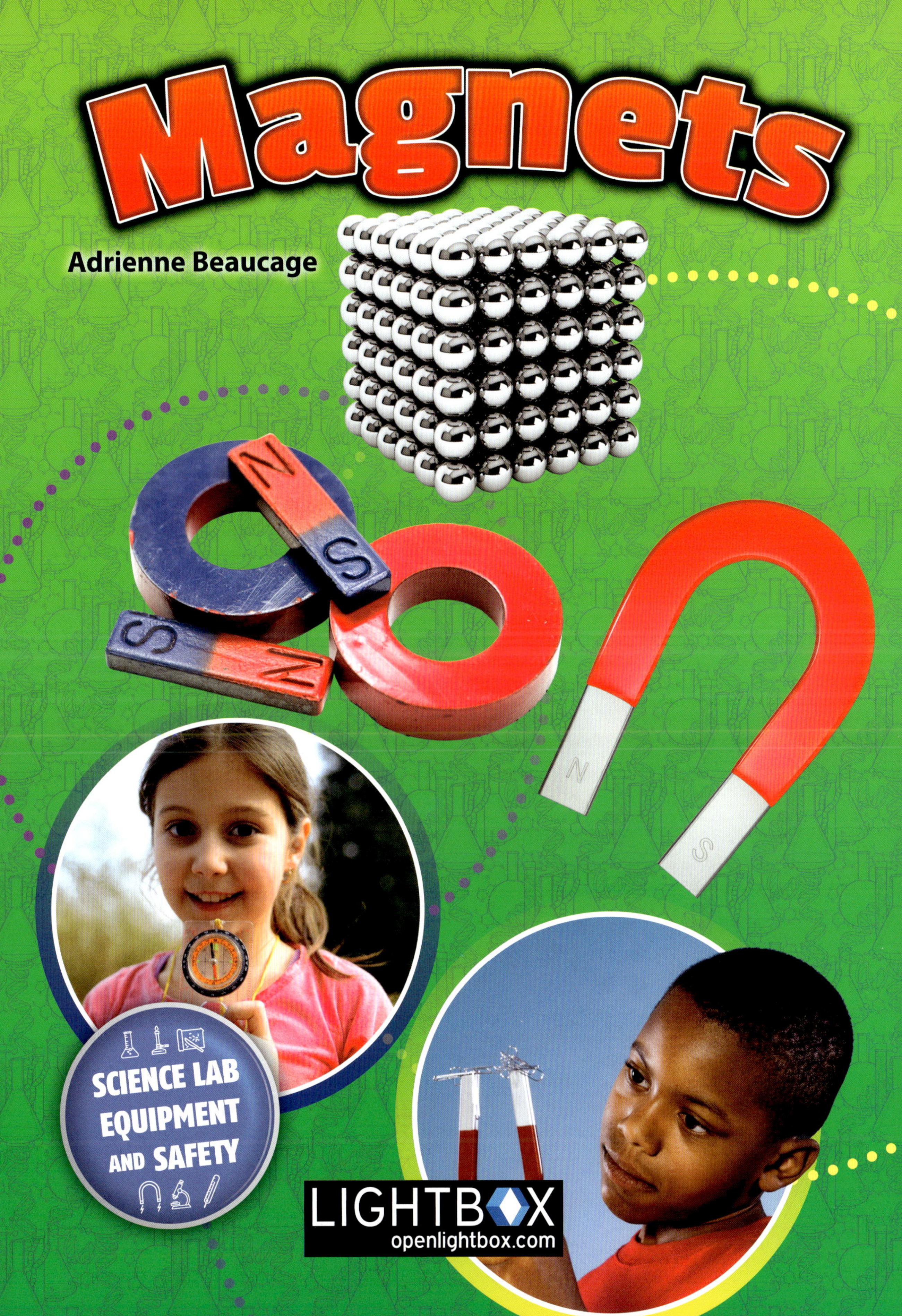
Magnets
Adrienne Beaucage
N
S
S
N
N
S
SCIENCE LAB
EQUIPMENT
AND SAFETY
LIGHTBOX
openlightbox.com

Go to **www.openlightbox.com** and enter this book's unique code.

ACCESS CODE

LBXW7456

Lightbox is an all-inclusive digital solution for the teaching and learning of curriculum topics in an original, groundbreaking way. Lightbox is based on National Curriculum Standards.

LIGHTBOX SUPPLEMENTARY RESOURCES

SHARE Share titles within your Learning Management System (LMS) or Library Circulation System

CURRICULUM Find national and state curriculum correlations

CITATION Create bibliographical references following the Chicago Manual of Style

STANDARD FEATURES OF LIGHTBOX

AUDIO High-quality narration using text-to-speech system

ACTIVITIES Printable PDFs that can be emailed and graded

SLIDESHOWS Pictorial overviews of key concepts

VIDEOS Embedded high-definition video clips

WEBLINKS Curated links to external, child-safe resources

TRANSPARENCIES Step-by-step layering of maps, diagrams, charts, and timelines

INTERACTIVE MAPS Interactive maps and aerial satellite imagery

QUIZZES Ten multiple-choice questions that are automatically graded and emailed for teacher assessment

KEY WORDS Matching key concepts to their definitions

Lightbox Grades 3–5 Subscription
ISBN 978-1-5105-5424-5

Access hundreds of Lightbox titles with our digital subscription. Sign up for a **FREE** subscription trial at **www.openlightbox.com/trial**

Magnets

CONTENTS

A FORCE OF NATURE

Science helps people learn about the world around them through **experiments** and **observation**. Sometimes, experiments lead to new discoveries or **inventions**. Have you ever wanted to do an experiment of your own? Maybe you can use the science laboratory, or lab, in your school. A science lab is a special room with **equipment** such as tools and safety gear.

Magnets are often found in science labs. A magnet is a rock or piece of metal that creates a magnetic field. This is an invisible field that lets magnets pull certain types of metal toward themselves. Some science labs have very large, powerful magnets. The magnets in your school's science lab are probably smaller ones that you can hold in your hand. Magnets can be used to **conduct** exciting science experiments.

Scientists have discovered **10 powerful magnetic stars**, called **magnetars**, in our galaxy.

Earth is a **giant magnet** with a field that extends more than **40,000 miles** (65,000 kilometers) from the planet's surface.

HISTORY OF MAGNETS

The effects of magnets were first observed more than 2,000 years ago. It is believed that the ancient Greeks were the first to use a **mineral** called lodestone. It would **attract** other pieces of the same mineral. The ancient Greeks called it a magnet.

William Gilbert is often called the "father of magnetism."

Magnets have been used for scientific purposes since the 1500s. An English scientist named William Gilbert was the first person to study magnets using the scientific method. Soon after, other scientists did experiments with magnets and learned more about their fields. People's understanding of magnetism increased greatly in the 1900s. This led to the development of new technologies. Today, much of what people know about magnets comes from the work of two twentieth century German scientists, Werner Heisenberg and Ernst Ising.

Magnets Timeline

1600
William Gilbert experiments with magnets and finds that Earth itself is a magnet.

1820
A scientist named Hans Christian Oersted learns how magnets and electricity work together.

1945
Engineer Percy Spencer invents the microwave oven, which uses electric and magnetic fields to heat food.

2019
Scientists discover a new type of magnet called a singlet-based magnet. Unlike other magnets, it creates magnetic fields that appear and disappear.

PARTS OF A MAGNET

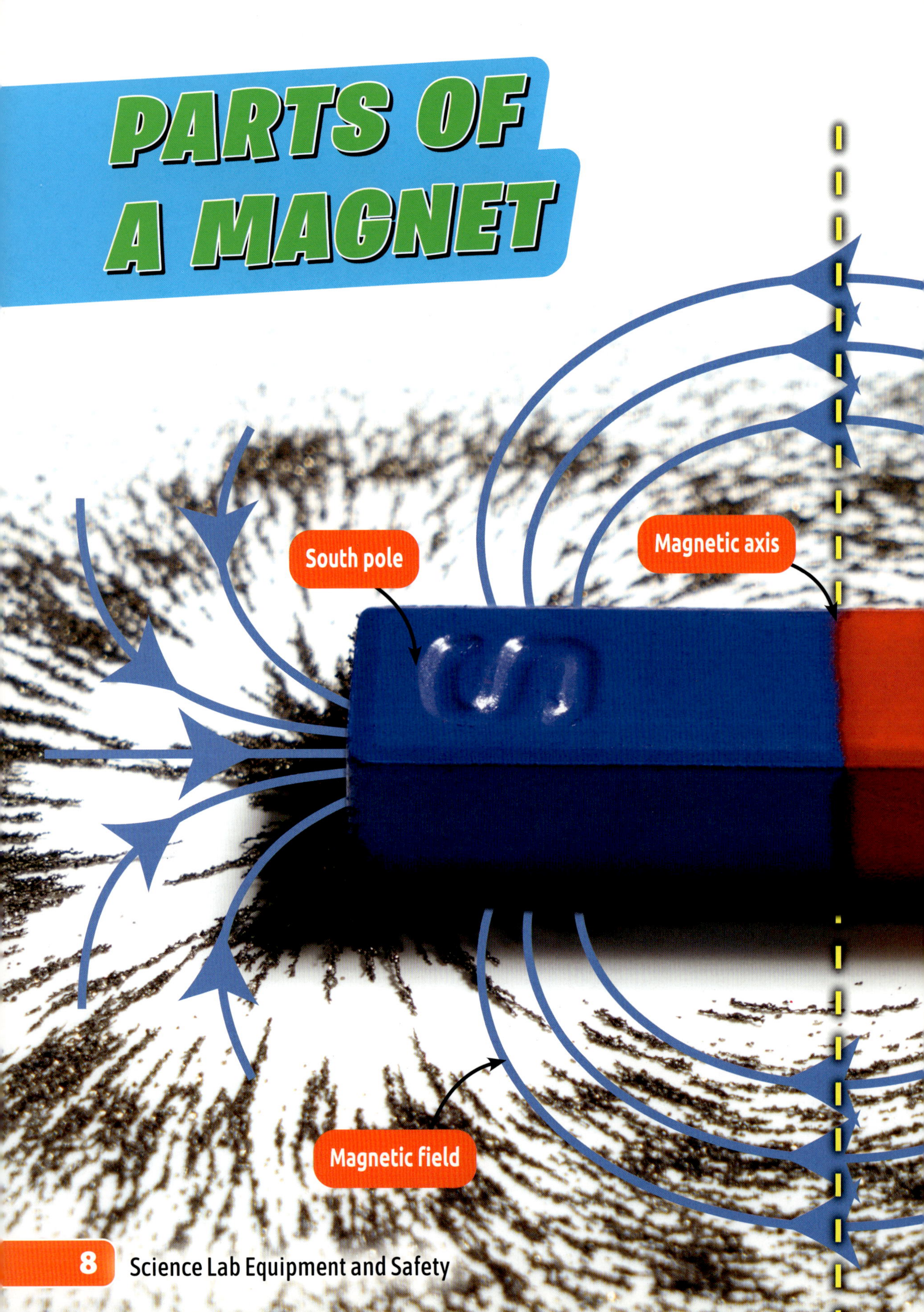

A magnet has two ends called poles. One end is the north pole, and the other end is the south pole. The magnetic force of a magnet flows from the north pole to the south pole, creating a magnetic field around the magnet. A straight line called the magnetic axis joins the north and south poles.

When two magnets come into contact, the north pole of one magnet attracts the south pole of the other magnet. However, the two north poles **repel** each other. The two south poles also repel each other. Similar poles repel one another, while opposite poles attract.

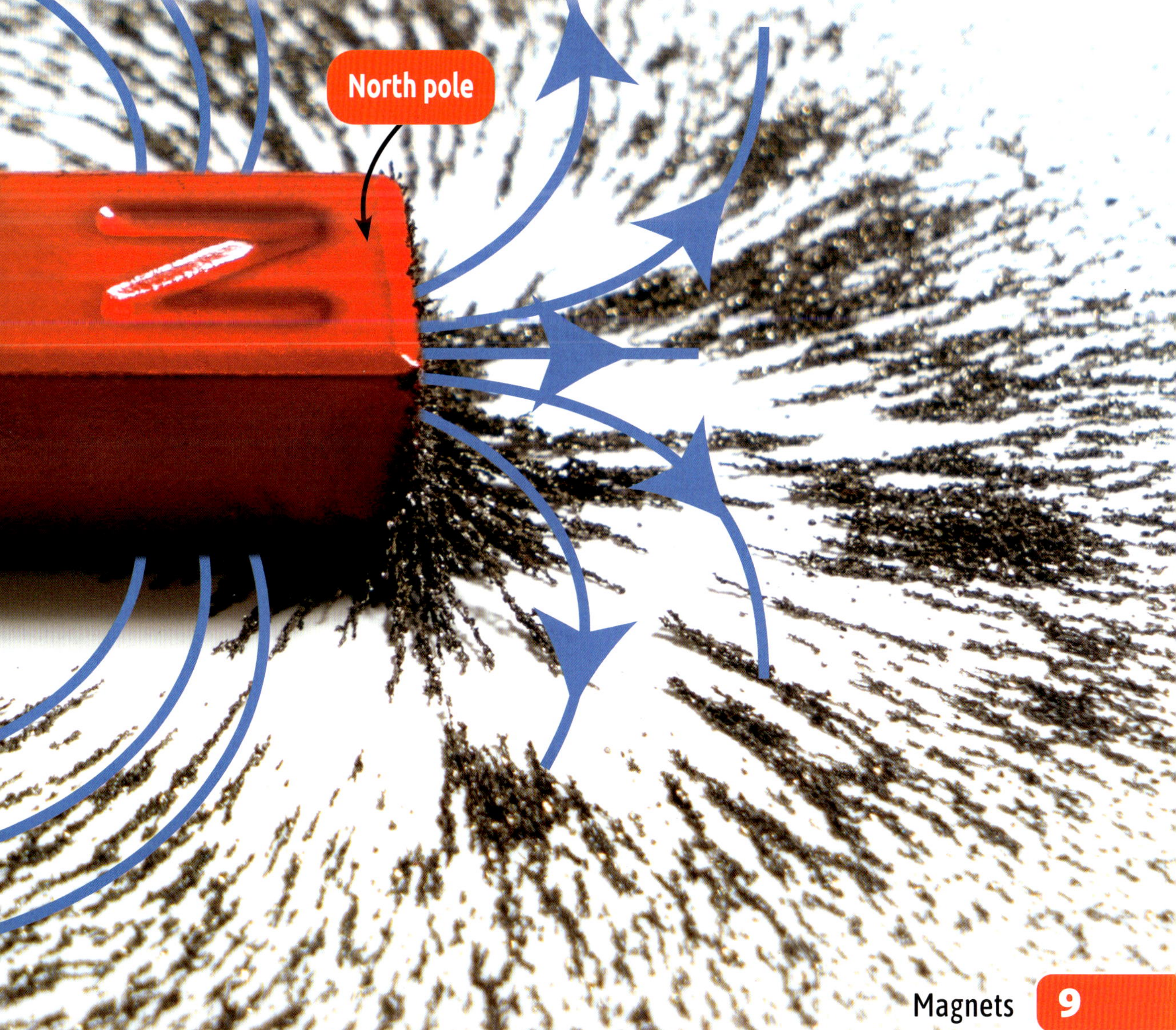

HOW IT WORKS

Magnets can be used to construct models.

Everything in the universe is made up of very small units known as atoms. An atom has a core called the nucleus in the middle. Tiny **particles** called **electrons** spin around the nucleus. These electrons carry an electric charge. As electrons move, their electric charge creates a magnetic field.

In certain metals, such as iron, nickel, and cobalt, most electrons spin in the same direction. The tiny magnetic fields from the electrons add up, making the metals magnetic. These metals are attracted to magnets, but are not yet magnets themselves. They can be changed into magnets by electricity or by other magnets.

Magnets and magnetic materials are everywhere. In addition to the magnets scientists use in labs, there are magnets in objects that people use every day. Many people put magnets up on refrigerators in their homes. Common objects such as fans, washing machines, cars, and even some trains use magnets to work.

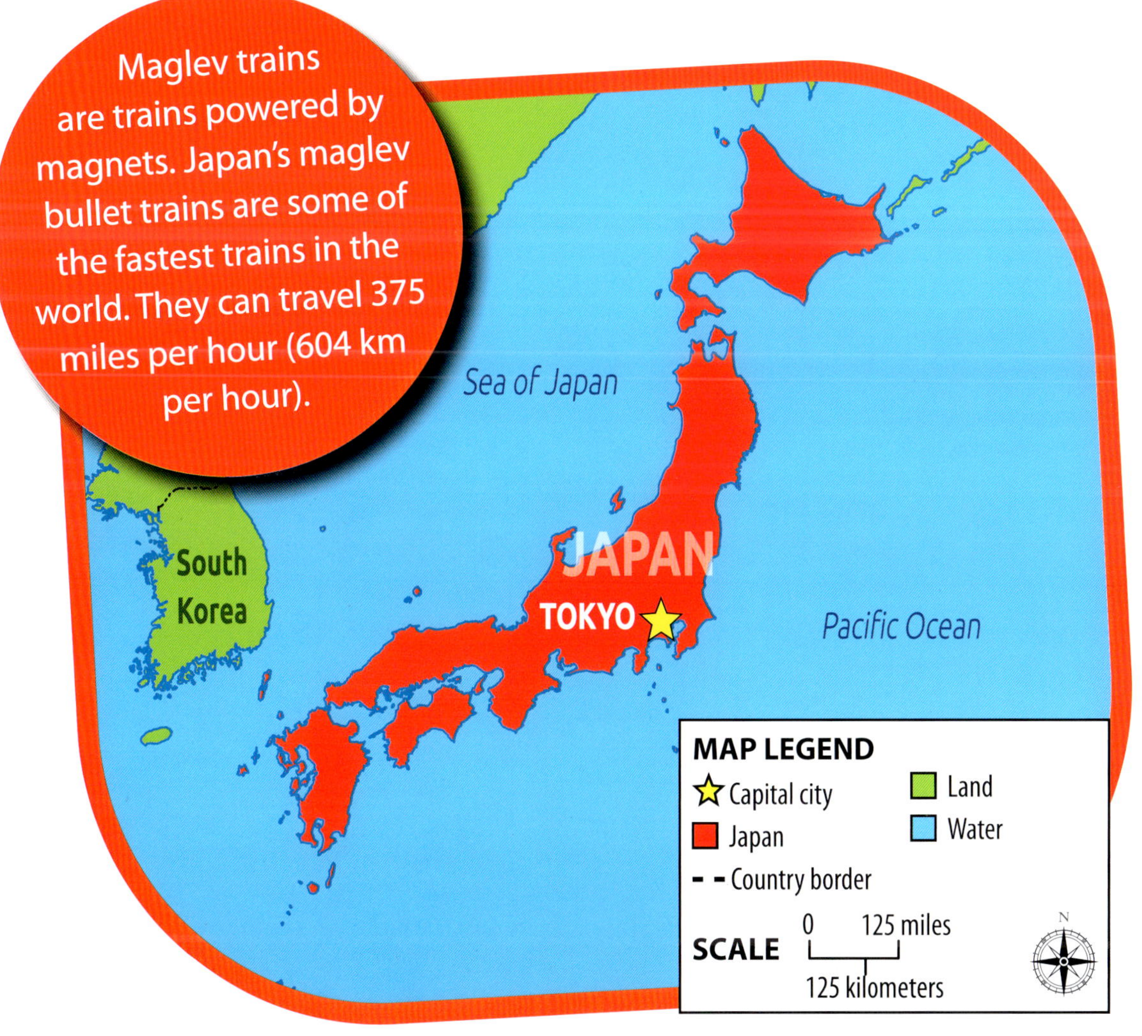

SAFETY IN THE LAB

Powerful magnets can damage electronics such as smartphones and watches. Avoid wearing a watch when handling magnets.

It is very important to follow safety rules in the science lab. Some tools and equipment can be dangerous if they are not used correctly. Learning the rules helps keep everyone in the lab safe.

Magnets should always be handled carefully. Even small magnets can be very strong, so it is important to use them properly. Magnets should never be placed in a person's mouth. Do not use teeth to pull magnets apart, and do not attach magnets to dental braces. It is very dangerous to swallow a magnet. If someone accidentally swallows a magnet, be sure to tell an adult right away.

Cutting a magnet **in half** creates **two new magnets**, each with its **own north** and **south poles.**

The **strongest** permanent magnets in the world are **rare-earth magnets**. They can attract **1,000 times** their own weight.

THE SCIENTIFIC METHOD

The scientific method helps people understand how the world works. Scientists use it as a way to answer questions. The method involves testing a hypothesis with an experiment.

1. QUESTION

Ask a question about something that you want to learn more about.

2. RESEARCH

Gather information about the topic. You can use books or the internet to do research.

3. HYPOTHESIS

Based on your research, write down what you think the answer to your question will be. This is your hypothesis.

4. EXPERIMENT

Design an experiment to test your hypothesis. Then, gather the materials you need and perform the experiment.

5. OBSERVATION

Observe your experiment and write down what happens.

6. ANALYSIS

Analyze your observations. Can you figure out what happened and why? This is called cause and effect. The cause is the reason something happened. The effect is what happened.

7. CONCLUSION

Based on your experiment and analysis, was your hypothesis correct? Write down your results and share what you learned with others.

EXPERIMENT

IS IT MAGNETIC?

Cassie is a young scientist. She is learning about magnets in school. Her teacher holds up a crayon, a paper clip, and an eraser. He asks Cassie's class if these objects are magnetic. Cassie thinks she knows the answer. Can she use science to prove it?

1. QUESTION

Are crayons, paper clips, and erasers magnetic?

2. RESEARCH

Cassie goes to her library and finds a book about magnets. She learns that some metal objects are magnetic. Cassie's school librarian helps her to research online, and she discovers that paper clips are usually made of metal. Cassie also learns that crayons are usually made of wax, while erasers are typically rubber. Based on her research, Cassie believes that only the paper clip is magnetic.

SAFETY FIRST!

Be careful not to put your fingers between a magnet and any object. If the object is magnetic, it may pinch your fingers when it is attracted to the magnet.

3. HYPOTHESIS

The paper clip is magnetic, but the crayon and eraser are not.

4. EXPERIMENT

Cassie uses a U-shaped magnet for her experiment. She places each of the objects on the desk in front of her. They are spaced far apart. One at a time, she slowly touches each object with the magnet.

5. OBSERVATION

The crayon and eraser do not move when they are touched by the magnet. The paper clip sticks to the magnet when touched.

6. ANALYSIS

The magnet generates an invisible force that attracts the metal paper clip. The effect caused by this field is the paper clip sticking to the magnet.

7. CONCLUSION

The experiment supports Cassie's hypothesis. The paper clip is magnetic. The crayon and eraser are not.

Only metal paper clips are attracted to magnets. Paper clips made of plastic are not magnetic.

EXPERIMENT

MAGNETS UNDERWATER

Have you ever clapped your hands underwater? Try clapping your hands in a swimming pool or bathtub. It is much harder than clapping your hands in the air. Jim wonders if water also makes it harder for a magnet to work. What can he learn using the scientific method?

1. QUESTION Does water impact how well a magnet works?

2. RESEARCH Jim knows that water is a liquid. He also knows that air is a gas. He uses the internet to research liquids and gases. Jim learns that particles in a liquid are close together, while particles in a gas are farther apart. He thinks the close particles in water may reduce the strength of a magnet's magnetic force.

3. HYPOTHESIS A magnet will have less force when used underwater.

SAFETY FIRST!

Avoid using hot water when filling the plastic tub. If the water is too hot, it could burn you. Cold water is safer.

4. EXPERIMENT Jim places a plastic ruler on a table and puts a metal paper clip beside it. He slowly slides a magnet along the ruler, moving closer to the magnet. The magnet attracts the paper clip when they are 2 inches (5 centimeters) apart. Then, Jim fills a plastic tub with water and repeats the same steps. This time, the ruler, magnet, and paper clip are underwater.

5. OBSERVATION Jim observes that the magnet attracts the paper clip underwater when they are 2 inches (5 cm) apart.

6. ANALYSIS Jim's experiment had the same results in the air and underwater. The water did not have any effect on the magnet's force.

7. CONCLUSION The experiment **disproves** the hypothesis. Water does not impact the force of a magnet.

A magnetic field is strongest at a magnet's north and south poles.

ACTIVITY

CREATE A COMPASS

A compass is a tool used to determine direction. It has a magnetic needle that points to Earth's **North Pole**. This means that a compass always points north. Follow these steps to create your own compass.

MATERIALS

- Bowl of water
- Bar magnet
- Metal sewing needle
- Red marker
- Sheet of wax paper
- Tape
- Scissors

STEPS

1. Rub the north pole of your bar magnet on one end of your needle. Only rub in one direction. Do this 10 times.

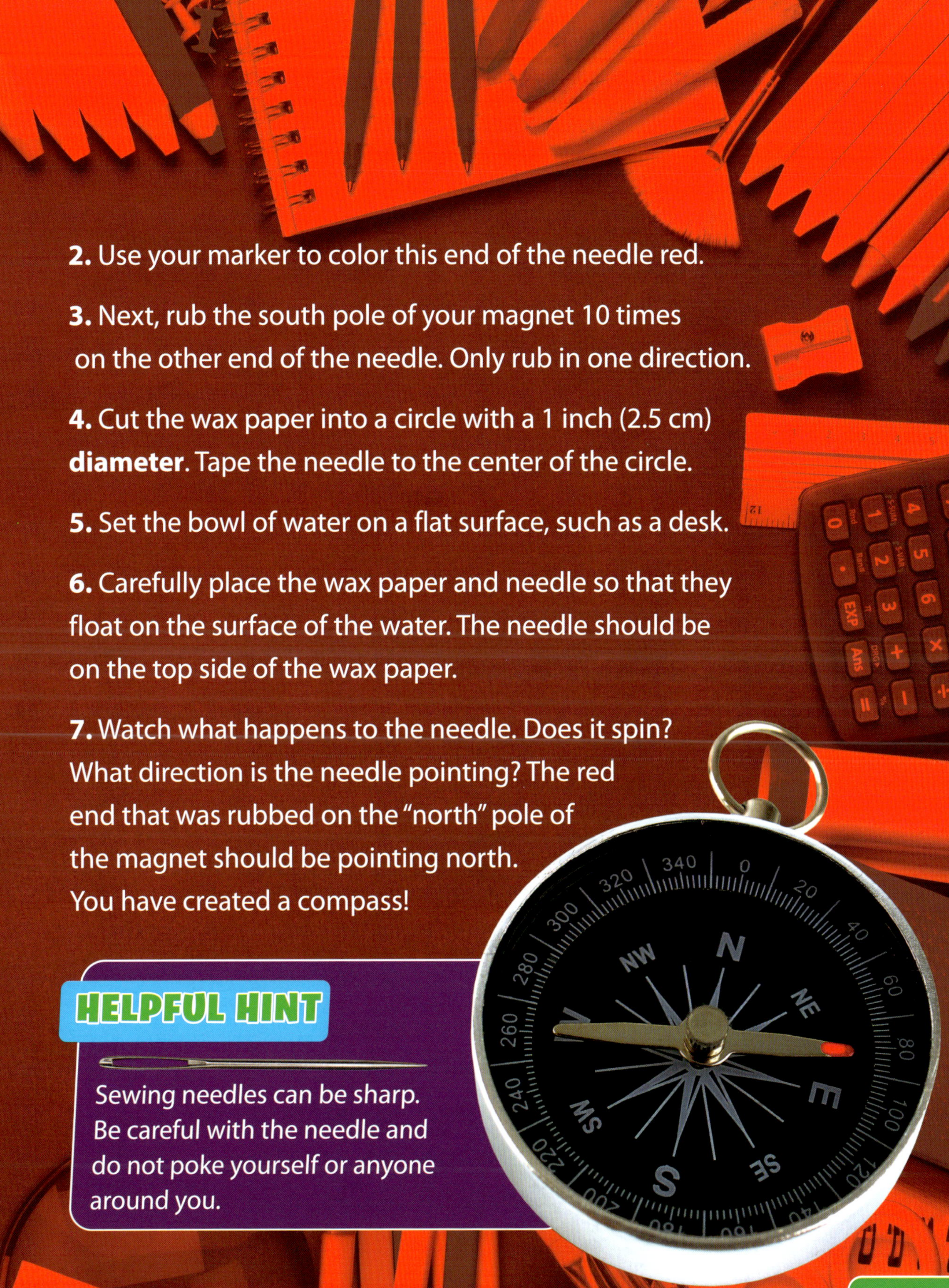

2. Use your marker to color this end of the needle red.

3. Next, rub the south pole of your magnet 10 times on the other end of the needle. Only rub in one direction.

4. Cut the wax paper into a circle with a 1 inch (2.5 cm) **diameter**. Tape the needle to the center of the circle.

5. Set the bowl of water on a flat surface, such as a desk.

6. Carefully place the wax paper and needle so that they float on the surface of the water. The needle should be on the top side of the wax paper.

7. Watch what happens to the needle. Does it spin? What direction is the needle pointing? The red end that was rubbed on the "north" pole of the magnet should be pointing north. You have created a compass!

HELPFUL HINT

Sewing needles can be sharp. Be careful with the needle and do not poke yourself or anyone around you.

QUIZ

1 How far does Earth's magnetic field extend?

2 What magnetic mineral did the ancient Greeks use?

3 Who first researched magnets using the scientific method?

4 What does cutting a magnet in half do?

5 How many poles does a magnet have?

6 Can magnets be dangerous?

7 What are trains powered by magnets called?

8 How can some metals be changed into magnets?

9 Who invented the microwave oven?

10 Is a crayon magnetic?

Answers:
1. More than 40,000 miles (65,000 km) from Earth's surface **2.** Lodestone **3.** William Gilbert **4.** Creates two new magnets **5.** Two **6.** Yes **7.** Maglev trains **8.** By electricity or by other magnets **9.** Percy Spencer **10.** No

KEY WORDS

attract: pull toward

conduct: carry out

diameter: distance across a circle through its center

disproves: shows that something is false

electrons: small, negatively charged particles in an atom

equipment: items needed for a particular activity

experiments: scientific procedures performed to discover something

inventions: useful devices or processes that have been created for the first time

mineral: a solid, non-living substance found in nature

North Pole: the magnetic south pole near the northernmost point on Earth's surface

observation: watching and noticing what happens

particles: small pieces of matter

repel: push away

INDEX

LIGHTBOX

SUPPLEMENTARY RESOURCES

Click on the plus icon found in the bottom left corner of each spread to open additional teacher resources.

- Download and print the book's quizzes and activities
- Access curriculum correlations
- Explore additional web applications that enhance the Lightbox experience

LIGHTBOX DIGITAL TITLES
Packed full of integrated media

VIDEOS

INTERACTIVE MAPS

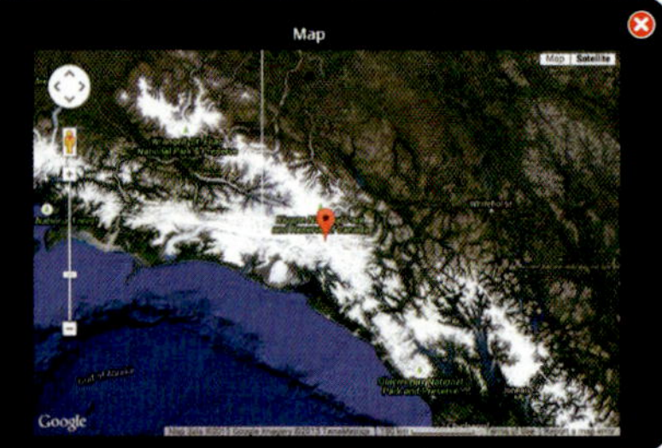

WEBLINKS

SLIDESHOWS

QUIZZES

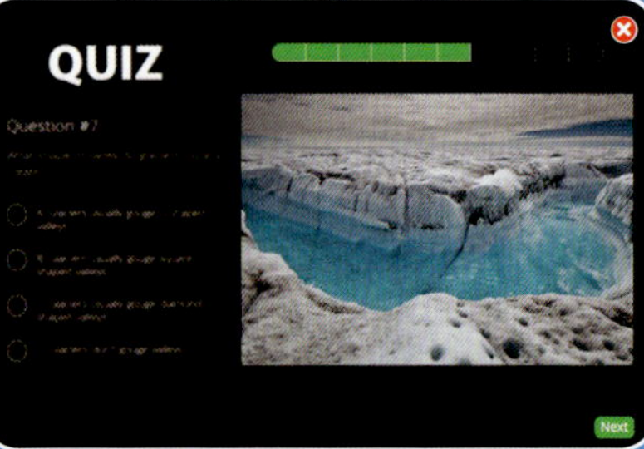

OPTIMIZED FOR

- ✔ TABLETS
- ✔ WHITEBOARDS
- ✔ COMPUTERS
- ✔ AND MUCH MORE!

Published by Lightbox Learning
276 5th Avenue, Suite 704 #917
New York, NY 10001
Website: www.openlightbox.com

Library of Congress Cataloging-in-Publication Data

Names: Beaucage, Adrienne, author.
Title: Magnets / Adrienne Beaucage.
Description: New York, NY : Smartbook Media Inc., [2022] | Series: Science lab equipment and safety | Includes index. | Audience: Grades 2-3
Identifiers: LCCN 2021032101 (print) | LCCN 2021032102 (ebook) | ISBN 9781510559080 (library binding) | ISBN 9781510559097
Subjects: LCSH: Magnets--Juvenile literature.
Classification: LCC QC757.5 .B43 2022 (print) | LCC QC757.5 (ebook) | DDC 538/.4--dc23
LC record available at https://lccn.loc.gov/2021032101
LC ebook record available at https://lccn.loc.gov/2021032102

Printed in Guangzhou, China
1 2 3 4 5 6 7 8 9 0 25 24 23 22 21

092021
111020

Project Coordinator: Priyanka Das **Designer:** Ana María Vidal

Every reasonable effort has been made to trace ownership and to obtain permission to reprint copyright material. The publisher would be pleased to have any errors or omissions brought to its attention so that they may be corrected in subsequent printings. The publisher acknowledges Alamy, Getty Images, and Shutterstock as its primary image suppliers for this title.